Handbook on DTH Transmission & communication

Television and Telecom

by

Chandra Bhushan S. Mishra
M. Sc., DSIM, PPC

1

Préface

While working on the project for new earth station I found what steps require to establish a satellite ink for telecom services. How much time it consumes to complete installation and finish testing with satellite operator. Later obtain certification from satellite operator to access satellite and operate the services.

Construction of PAD for antenna, transmission or equipment room to accommodate GCE and related equipment in racks and hand over this to vendor for installation.

Arrival of antenna, GCE, and a lot of connector and cables which includes power cable to the site. Assembling antenna support, reflector, and local motor controller cabling to powerup AZ/ EL/pol motors etc.

GCE rack fixing and placing equipment in it to their designated places, power cabling and integrating system in racks to complete transmission and reception chains, configuring equipment(s) and testing.

Finally carry out in-station test and performing test with CSM allotted by Satellite operator, as per their given guidelines. Once test is passed satellite operator will permit to start the services.

I was involved in Telecom and television satellite system from initial installation to operation of services and maintenance of the antenna and associated equipment. Friend requested me to put in writing this so that it may be beneficial to others working in this field or wanted to know how it works.

This book is submitted to you for brief introduction both the system telecommunication and Television transmission on D2H platform. Hope you will find it useful.

Chandra Bhushan Mishra

Index

Chapter-1

<u>Documentations</u>

Before an earth station established there are lots of field and paperwork to be done. The most important of them are proposal, site survey, and estimation. An organization plans to have an earth station must get approval from the ministry concerned (most of the countries consults ministry for communications for this. In some countries, this responsibility has been given to ministry for transport). Once organization gets ministerial approval it invites global tender for establishing earth station. Vendors may ask questions related to this project. Vendors may request a meeting with technical team of the organization to understand requirement in better way before submitting their quotation for the project.

After getting quotations from multiple vendors Technical and commercial team together will evaluate suitable vendor. They may ask vendors on some unclear points in their quote. Finally, vendor, whose tender is honored, will be informed to start kick off meeting and submit work scheduled, completion plan etc.

A team from the vendor's company will visit the organization to get further information such as site selection, approach road to the site and visibility of satellite which the organization want to access etc. This team is responsible for time frame of the work. All necessary questions, technical and administrative, team leader of vendor will ask from the organization's staff responsible for this project. In-fact this team of the experts consists of technical and administrative personalities. Organization may grant technical management services to the vendor. In that case Vendor may ask organization, coverage area of the service, type of the service, satellite to be access etc. Then Vendor will study suitable satellite for delivery of the service. But it is very rare. Generally, organizations have already planned it. They already have in mind-

- Satellite to be accessed
- Frequency band to be used C, Ku, DBS, or ka Band.
- Transponder in which service will be present.
- Service to be offered- Voice. Data or Video.
- Geographical coverage area of interest.

WORK SCHEDULE DOCCUMENT:

Once this team left the place, they will send the documents pertaining to the work. This document is known as Project plan [with details of timeline]. These details as follow: -

1. Date the work is scheduled to start.
2. Date of the foundation to be ready as per diagram.
3. Date of the shipment made and arrival at site.
4. Date of the installation team will arrive.
5. Date on installation will begin- casting of PAD.
6. Date on antenna installation start.
7. Date on Ground communication system will install (Generally, items 6, 7 may starts together to finish project quick)
8. Dates showing each phase, when to complete.

9. Date on in-station test will start and complete.
10. Date on antenna verification test start and commissioned.
11. After passing antenna approval test, earth station will receive access permission from satellite operator to access their space craft.

Approval of the Project and its timeline will be given to vendor. Vendor or company will send the standard foundation chart for antenna pad and as per diagram organization may sub-contract civil engineering firm to get foundation ready. If you have left this job to Vendor, then they will send civil engineering team to get this job done. Once the foundation is ready, they will ship Antenna material to the site with mechanical engineering team. This team will install antenna. Note foundation needs at least 15 days to cure properly.

Depend upon agreement on project plan electronic engineer may also come together with mechanical team to install GCE to speed up project deliver. This will give mutual understanding within team and project finishes quick.

EQUIPMENT RELATED DOCCUMENTS:

Now the vendor will send ground communication equipment along with
1. Installation manual.
2. Operation manual of all equipment.
3. Factory acceptance test data sheet of all equipment
4. Warranty cards and shipment address to send faulty equipment.

5. Station equipment configuration data sheet of each equipment.
6. System level diagram.

The above document you will get in duplicate or triplicate as per negotiation with company and organization during initial contract. You should keep these documents at secured place. These may be required during preventive maintenance for comparison.

ON SITE DOCCUMENTS:

Engineering team should handover following documents to team looking on Operations. These are very important documents for earth station in-charge/ operator. This gives you idea in case trouble shooting on problem in earth station. You will get most of the critical data in these documents, which is required in the case of emergencies.

1. Factory test data sheet for the equipment's.
2. In-station test results.
3. Configuration and setup details of equipments.
4. Level Diagram of the uplink link and down link chain.
5. Final testing and commissioning results. (SSOG)
6. Station Antenna certification from satellite operators.
7. Operation and Maintenance manual for earth station equipments in general.
8. List of the spares supplied by installation team. (This is a complex issue. Spares will be depending upon initial negotiation between both the administrations)

9. Contact person(s) in case of problem after hand over of the project, their telephone, fax number and e-mail address.
[This required for help during problem, particularly within the warranty period]

An example of approximate time schedule is given in Appendix- A.
This is just an idea not exact schedule. Actual Schedule depend upon size of the project, efficient deliveries of the equipment by manufacturer, number of persons involve and efficiency of the work force and of course payments on the project.

Normally manufacturer needs minimum 12-15 weeks to supply equipment after receiving firm order or payments. So, you can imagine if the project is small, say only one antenna. It will take minimum 5 to 6 months to come in operation while big projects may take 1-2 year or even more to complete all while one by one operation may start from 5-6 months.

Schedule for Earth Station Installation & Commissioning. Please see Appendix –A. for details format and Documents.

Chapter-2

In this chapter, you will learn function of:

1.   How earth station works.
2.   Modulator
3.   Up Converter.
4.   High Power Amplifier.
5.   LNA and LNB
6.   Block Down Converter
7.   Integrated Digital Receiver
8.   Antenna Tracking Control unit

After completion of this chapter, you will fully be aware of functioning of equipment in the communication link. You will be able to judge the problem in chain in case it arrives.

How Earth Station works:

Earth station is very important when you want to send your signal overseas or within the country where cable connectivity is not available. This avoids use of the copper or fiber cable to send your signal long distance. Since this work on RF so satellite coverage area is only limit. The signal can be down linked anywhere within its coverage area via suitable dish antenna. Its size depends upon the frequency band of the signal to be received and down link e.i.r.p. of the satellite transponder in that area.

Initial cost of getting earth station is little high, but maintenance cost is very little compare to other transmission systems. This provides un-interrupted services. There is very little chance for break down. This gives excellent quality of service throughout year. Sun satellite interference is the only known phenomenon. This affect satellite services twice a year. This affected services for 2 to 10 minutes for 5 to 6 days. This affects different satellites at different time depends upon its longitude. Other degradation of the service may be due to heavy rain in DBS, Ku, and Ka band while C-band remains un-affected or affected very less.

Your signal will pass through ionospheric region and is received by satellite in frequency range of GHz. The setup is split into following groups.

    i.   C- Band (6 GHz up and 4 GHz Down)
    ii.   Ku-Band (14 GHZ Up -12 GHz Down)
    iii.   DBS Band (18 GHZ Up 12 GHz down)
    iv.   Ka Band (25 GHz Up 12 GHz Down)

These are used for satellite telecommunications and Broadcast DTH TV signal feeds. Frequency given above just to give an idea of band on satellite. Specific frequency plan can be had from satellite administration in required band.

Let me explain how earth station is used in both the cases: Telecommunication and Television.

A. Telecommunications Earth station.

Now analog system is outdated so I am going to explain you about digital transmission only. Voice signals from exchange reaches earth station in E1 (2 MB) or T1 (1.5 MB) stream. Here this E1/T1 stream is further compressed in ratio of 4:1 (now day's equipment is available for compression ratio 8:1 operating successfully. Thus, you can compress 120 or 240 -voice channel to form a packetized carrier of 2-MBPS. DCME and IAT equipment are used to get this compression and de compression in voice channels. The compressed voice signals further passthrough IDR modem where Forward Error Correction (EFC and overhead signals are added to it.) Signal in the form of IF (70 MHz or 140 MHz) from the modem pass to up converter to convert IF into RF carrier required for transmission. The RF power of the carrier from up-converter is not enough to illuminate the satellite. Hence, High Power Amplifier is used to give required output power to the carrier. Generally, 500/ 750 watts HPA is used in this case. HPA out put is fed to antenna through waveguide and feeder system for transmission of the signal towards satellite.

The same antenna is used for transmission and reception both. Antenna can have two port or four port feed systems, Horizontal and Vertical in transmission and reception. This makes easy for uplinking in any one or both the polarization systems namely Vertical and Horizontal.

Similarly, Low Noise Amplifier (LNA) amplifies received weak signal, which is catch by dish antenna. Here signal gets amplified more compare to noise associated with it. RF Signal then fed to down converter to get IF from received signal. This IF signal, 70 or 140 MHz, depending upon selection at transmit station. Further IF signal is fed to Modem which separtes necessary signals from this IF and feed E1 /T1 digital stream to de-compressing equipment. Here 1:4 or 1:8 decompressions achieved in individual time slot is accomplished from 2 or 1.5 MBPS normal stream. This further passes to digital switch system as a telephone channel of size 64 kbps, which carry voice information's. Thus, on one 64 kbps channel 4 or 8 voice information is passed. Speech interpolation technique is used to achieve this.

DATA channels are passes in both the directions as clear channel. This provision is available in the compression equipment. DATA channel MUST NOT be compressed as in case of voice channel we have.

A block diagram is given in the Figure-1.

B. For Digital TV Transmission.

The Audio Video signals from studios are received at earth station. Here we do amplification and distribution of audio and video signals in ADA (Audio Distribution Amplifier) and VDA Video Distribution Amplifier) respectively. A/V signal is fed to encoder and Mux for further transmission.

Now in the digital platform Audio and Video signal are embedded and output is SD-SDI/ HD-SDI signal. This SDI signal is fed to Encoder (digital compression equipment) Encoded ASI signal is fed to Multiplexer, in case of MCPC mode of transmission, which has capability to re-multiplexed 15-18 digital channels (even more) to form a Transport Stream for further transmission. As shown in the diagram Fifure-2.

In the case of SCPC, digitally encoded signal, ASI, is fed directly to modulator for further transmission. Thus, we can transmit more TV channels in the limited bandwidth on satellite transponder. In comparison to analog signal, where a TV channel needs at least 18 MHz bandwidth space on satellite transponder, Digital TV channel, in SCPC mode, takes very little space (typically 3 to 4 MHz) on satellite transponder. Thus, satellite can accommodate more TV channels on a transponder. Hence increase in capacity to carry more TV channels on satellite and simultaneously low cost to the uplink services. In other words, you can transmit more channels at cost of one analog TV channel. How economic digital system is?

14

A multiplexed TV signal, which is known as Transport Stream (TS), this ASI signal is on passes through modulator, which confined these signals in the form of IF (140 MHz). The IF signal further passes through Up Converter. Here IF signal get converted into RF carrier frequency, which is required by satellite for the transmission. Since the power of output of RF carrier from Up Convertor is not enough to illuminate the satellite hence RF power is boosted in the High-Power Amplifier. Out put power of HPA is quite good and is adjusted as per requirement by satellite operators. Even keeping power margin to compensate in case of rain etc. {750 to 3.0 KW power from HPA}

Down link RF signal, which is send down from satellite is received by Dish antenna and fed to Low Noise Amplifier (LNA) or low noise block converter (LNB). If you are using LNA you need BDC in the chain to convert RF down link to L-Band signal (See Figure-3) while LNB receives either in C band signal or Ku band signals {depends upon system in use} convert it to L-band in the range of 950 MHz to 2150 MHz. This signal is fed to Integrated Receiver Decoder (IRD). To get required TV signal for viewing or monitoring. Professional IRD are used in the technical facilities while commercial IRD are used by people for watching TV channels at home. These IRD have options to tune or search TV channels and store in the memory. You may recall any channel and view it on the TV monitor.

The antenna must be all time look to satellite, for this purpose Beacon signals are radiated by satellite Operators. This beacon signal is received by beacon receiver. Beacon receiver translates it into DC voltage corresponding to the received signal strength. This DC voltage is fed to Antenna Controller Unit to peaks the antenna to given satellite. Once voltage goes down from the set value, ACU become active and start tracking for the satellite. Once ACU peaks satellite signal strength of beacon gets increased hence corresponding DC output voltage from the Beacon receiver increases. This goes to ACU where it is compared to benchmark voltage and ACU stop tracking at the best value found during tracking. Thus, antenna and satellite all time remains in the line of site

This is simplified description on an earth station and its functions. You understand how an Earth station works by now. This is wonderful? Isn't it? Your signal from antenna radiates in electromagnetic form and reaches satellite, which is 38,000km above earth. Satellite translates and sends it back to the ground.

Antenna receives this electromagnetic signal and converts into electric signal by LNA/ LNB. Here signal gets amplified. In LNB either you fed C- Band or Ku- Band signals, depends upon your LNB band, you will get output L- Band in the range of 950-2150 MHz So you must decide what frequency band you are receiving from Satellite dish and hence you must choose LNB accordingly e.g., for C- band, Ku, DBS, or Ka-band.

Different types of the LNB available in the market. You can select one, which suits your requirement. There are multi feed LNB which receive signals from dishes in C, Ku, or DBS band. People are using three or four antennas at home all fitted with individual LNBs. You can use DiSEqC Switch which takes 4 to 8 LNB signal input and gives output at one port.

Some LNB's accepting one input and giving 4 output for different set ups. Four outputs are H- high and low and V- high and low. While in some LNB's you may get both polarization on a single cable. These LNB's are called single solution LNB generally in C-Band for customer use at home.

LNB out put is fed to Integrated Receiver Decoder (IRD) and Audio Video out put taken to TV Monitor for viewing purpose. You must select transponder frequency and satellite you want to watch in the IRD. IRD/ STB will search that program for you. If you are using single solution LNB then care must be taken for entering LO frequency for H and Vertical as given on the LNB.

IRD's are also available mainly two types namely professional and commercial. Professional IRD is used by Earth Station and Cable operators while commercials are for customers use for receiving DTH signals. Professionals IRD are quite costly and have built in monitoring facilities while commercial is cheep and not highly informative.

Note: Descriptions of the individual equipment is beyond scope of this book. Only you will get idea what is does. For further details on equipment, you should go for another book by Author.

17

Modulator:

Digital TV services uses DVB-S and DVB-S2 modulators depend upon requirement of modulation mode:
Transport Stream (ASI) from the multiplexer connected to ASI input port of the modulator. For given bit rate modulator is set as follows: -

Parameters	Value set	Value set
Data set	ASI-1	ASI-1
PKT Size	188 bytes	188 bytes
Error Flag	Off	Off
Bit Rate	38.01476mbps	38.01476mbps
Symbol Rate	27.50 MSPS	27.50 MSPS
Coding	DVB-S	DVB-S
Mode	QPSK	QPSK
FEC	¾	¾
Roll Off	20%	20%
IF Frequency	70.0 MHz	70.0 MHz
IF Power	- 15 dBm	- 15 dBm
IF Out Put	ON	ON
Mod State	ON	ON
IF Spectrum	Normal	Normal
IF low Power	- 20 dBm	- 20 dBm
Carrier Act	Normal Power	Normal Power
Power Mode	Active	Active
Save Configuration: save at position: 0, 1 or 2 as convenient to operational team.		

In the modulator, you can set coding, symbol rate, bit rate, FEC, IF freq and power to suit your system requirement.

- Coding can be set to DVB-S or DVB- S2
- Available Modes are QPSK, 8PSK, for satellite.
- FEC can be ½, ¾, 5/6, 7/8 etc.
- IF can be either 70 or 140 MHz
- IF Power level to be set as per level diagram of the system. The range is from -10 to -20 dB.

Symbol rate and Bit rate are interrelated. Once you set symbol rate, bit rate appears automatically. Thus, same bit rate to be set in multiplexer to give correct matching otherwise modulator will give alarm – mismatch bit rate.

Up Converter

IF from the modulator is fed to up converter. Here IF signal get amplified and pass-through 1st stage of Mixer. the Reference oscillator 5/10 MHz used to generate LO frequency to mixed with IF from Modulator and it gives 1150 MHz 1st IF frequency. The 1st IF again is filtered out, amplified, and then mixed with synthesizer frequency, SHF*, (2780 to 13280 MHZ) to give correct RF frequency. RF is filtered amplified and then passes to HPA. Basic Block diagram of the up convertor is given on next page. This will explain different stages inside up convertor and how signal flow from it. This is brief explanation of up converter. See Figure-4

Note: * SHF frequency band sensitive, it is different for C-band, Ku, DBS, and Ka bands.

Chapter- 3

This chapter is described function of high-power amplifier, its working and inside component.
There are three types of HPAs

1. Solid State Power Amplifier
2. Travelling Wave Tube Amplifier
3. Klystron Power Amplifier

Details:

1. Solid State Power Amplifier: This type of power amplifier is in use where low power transmission is required 20 -50 Watts. Such as VSAT, 2-MHZ telecom services transmission e.t.c. It has advantage to install out-side near dish, no cooling system require, air cooling is enough.

2. TWTA is used for medium power transmission system. It can give power up to 750 watts. Multi carrier transmission system generally uses TWT amplifier.

3. KPA: Klystron power amplifier is generally used for Video transmission over full transponder. It comes from 2.0, 2.5, 3.0 KW power. System operator may get link budget from satellite operator and then decide which power range KPA is required to cater proper service for longer period.

<u>Working of HPA</u>:

1. RF from Up converter is fed to SSIPA
2. SSIPA has Attenuation control input to TWTA for proper RF amplification required for TWTA.
3. TWTA further amplify RF signal to get required EIRP by satellite operator.
4. A portion of RF out is taken though coupler for RF metering such as RF-out, RF reflect power, etc.
5. SSIPA is controlled by micro-processor controller for necessary signal adjustment.
6. All the adjust, display- mon, can be done from front panel or remote control.
7. Display system at front panel shows metering on selection of field.
8. KPA block diagram is given below in Figure-5 & 6 with TWTA, and KPA respectively.

Chapter – 4

In this chapter, you will learn:

1. Planning for preventive maintenance
2. What to observe every day and how observation is made
3. Carry out maintenance on periodic basis.

After completing this chapter, you will be able to make effective planning for maintenance of an earth station installations and Ground Communications Equipment (GCE).

Planning for maintenance

"Well planed is half done". Therefore, before doing any maintenance let us formulate a plan what maintenance to be done and when. This is to be decided once then it will be life long for the station. Be careful in fixing maintenance schedule. Indoor maintenance can be done any time but outdoor maintenance to be carried out in clear and clean weather. Do not carry out out-door maintenance during rainy season or when raining outside. Also, avoid doing maintenance if heavy wind with storm/ dust etc is present. This is to be decided considering country weather and seasons. Fix outdoor maintenance during Lukewarm season so that person does not feel hot while working on the antenna structure.

For an example, we propose maintenance plan as per attached sheet. This proposal may please be reviewed and get approved for action.

Please see attached sheet at the end of this chapter.

Part-I

Routine and Maintenance of Satellite Earth Station.

For proper working of the earth stations GCE, Power plant and antenna etc it is proposed to do following periodically.

Daily

1   Inspection of equipment in the chain- on line and confirmation of the same. In case any equipment came

online from the standby mode, then cause of switching to be investigated for failure of the earlier online.

2. Recording HPA details such as Output power, Beam voltage, Helix current, Collector current, reflected power and inside equipment temp etc.

3. Check Audio and video levels, quality coming from studios and down link if you are TV up link earth station.

4. HPA output and LNA/LNB output must observed continuously.

5. Spectrum Analyser to be hooked up in down link chain to see down link carrier level.

6. C/N and Eb/No. checked and recorded for comparison against benchmark. It may vary if atmospheric condition is not good.

WEEKLY

1. Change over of the configuration of the GEC chain and keep recorded. For example, suppose following is online chain in the earth station
Mod-1, UP Conv-1, Lin-1, HPA-A
Then, it should be changed to next week as follows
Mod-2, UP Conv-2, Lin-2, and HPA-B

2. Checking of Diesel Generator batteries voltage, water level and radiator water etc. Then start generator without load for testing purpose of the generator.

Monthly

1   Measurements of RF level for up and down chain and record the same. Modulator output power, Up Converter output power, HPA output power and LNB output. Of course, arrangements to be made for continuously monitoring of HPA and LNA / LNB output.

3.   Cleaning of filters of the GEC equipment to be carried out. HPA and power supply air intake filters to be cleaned. Encoder and Multiplex air input to be cleaned.

4.   UPS maintenance to be carried out and discharge test on station load to ensure backup capability and duration of the system.

5.   Generator to be run on load for one hour. Diesel to be checked and get filled if require.

Quarterly and Half Yearly

1   Antenna Lubrication and greasing to be done. Since the tracking movement is quite low being stationary satellite so lubrication frequency should be kept once in 3 months.

2   Antenna inspection to be carried out at the same time for loose and worn out, rusted and damaged parts. Antenna hub to be cleaned and inspected for damage, rusted parts. Cables and connectors to be touched for proper tightness.

3   Antenna dish to be washed so that dust and dirt washed away.

Yearly:

1. Change Engine oil in Diesel generator engine, change of oil filter water separator filters, checking and replacing belts if worn out or aging indication seen.
2. Checking for current supply of the DG starter Batteries.
3. Work out on antenna structure for rust removing and painting, earth resistance measuring, hub inspection, electrical terminals checking etc. washing of dish, checking IFL cables and tie it properly. Apart from regular lubrication of antenna. Gear oil of AZ /El motors, braking pad also checked annually.

Part-II

Ground Communications Equipment (GCE):

GCE being intermediate equipment's chain between Terrestrial link one side and satellite transmission on other side. Its vital role in the chain is quite interesting. It accepts Terrestrial signal, analogue audio and video, SD or HD embedded A/V signals, do necessary conversion, compression, encoding and multiplexing forming an ASI transport Stream, feeding to modulator to change TS to IF and then IF to Radio Frequency in upconverter and boosting its power required to illuminate the satellite for signal transmission through HPA.

New technology has made it possible to construct such equipment very simple by use of VLSI chips. Due to this no maintenance on the equipment required. If any maintenance is needed that is cleaning of the air vent filter, changing of the power supply setting. That is once you are replacing equipment with new one. Following maintenance is recommended of this equipment:

Daily:

1.   Just to have quick look on the *online Equipment* for abnormality, minor alarms, silent change over to Standby mode from online mode etc. Normally the equipment will not go in standby mode unless the equipment had experienced some problem. It may be for few seconds only.

2.   Take a note of the HPA readings such as- Beam Power, Collector Current, Helix Current, output power, Temp, Reflected Power and attenuator setting etc.

3.   Take a note on Audio level in put on each program. In case abnormality report to studios concerned. (If you are a TV earth station)

4.   You can measure level of DATA channel if you are telecom organisation.

WEEKLY:

To take online equipment chain to standby and standby equipment online manually.

This practice checks equipment health periodically. This enables Tech to take proper action well in advance. Some time equipment has changed its parameter due to fault in its electronics and not generated any alarm. So, you will presume that equipment is ok. When it goes online then you find it is not working. It is too late at that time because online equipment is faulted, and standby is also faulted at the same time. This causes break in service. It takes longer outage hours or (in days) unless you have any spare equipment in the stock to replace it (in hours).

 For Example:

Up converter's PLL has changed frequency or frequency drift occurred it will not give any alarm, but converters output frequency will be changed which is something else than your carrier causes anxiety, frustration and longer outage once on-line UP converter fails. This is one of the examples. Hence Recommended Configuration is as below: -

1.   HPA-A, UP-Conv-1, Modulator-1, Mux1 _Online_
        HPA-B, UP-Conv-2, Modulator-2, Mux-2 _Standby_

This is configuration for 1st week and so next week it should be changed like:

2.   HPA-B, UP-Conv-2, Modulator-2, Mux-2 _Online_
        HPA-A, UP-Conv-1, Modulator-1, Mux-1 _Standby_

Monthly:

1. Cleaning of the filter and equipment to be carried out. This should be done with Vacuum cleaner/ Blower and soft cloth soaked with water. This allows equipment's cooling system to work properly. Thus, longer lasting the equipment and less possibility to go faulty its electronics.
2. Measuring and recording levels on all the equipment and take a note. This enables to keep proper chain's level diagram which helps in level adjustment and detecting faults. Figure-7 & 8 (TV and Telecom system respectively)
3. Plot S.A. waveform of all equipment and keep for record. This will help in future adjustment and or repairs.

Annually:

1. Work on equipment for cleaning, touch up in case of the equipment chassis gone rusty or paints has gone off.
2. Touch each piece of the cable connectors (F, BNC, N, and SMA) and tighten it if has gone loose.
3. Check and tighten DB connector screws in each equipment or check latch if it is latched type connector.
4. Touch for firmness of power cord on all the equipment by pushing it. Do not pull power cards.
5. Tighten earth connector on each equipment.

6.   Measure frequencies and levels of the Ref Oscillator, PLL, on each equipment and IF, RF etc and compare with standard values. If necessary, do adjustment as specified in the equipment manual.

7.   Lubricate rack slide guide for smoothness.

Note:

It is recommended to open one by one equipment and see internal wiring/ connectors for damage, burn due to excessive heat and any other problems like salt formations itching out of the circuit board and PCB etc (in case Earth station is near coastal area).

This is particularly on HPA, which generates much heat inside. **Warning** to be strictly observed when working on **HPA**.

NOTE:

1. <u>Never carry out any maintenance on HPA alone</u>.

2. While working on circuit board, antistatic precautions to be taken.

3. Never open any circuit board from antistatic plastic unless you have wear earth strap on yourself. Serious damage could occur to the board.

See table-1 and Table-2 on schedule and maintenance records

Chapter- 5

In this chapter, you will learn how to carry out maintenance on
1. Antenna structure, motors, jackscrews, and bearings etc at lubrication points.
2. Ground Communication Equipment

After completion of this chapter, you will be able to supervise and or carry out meaningful maintenance by your team.

Antenna Lubrication:

To get optimum performance from satellite Antenna proper maintenance and care needed. Periodic lubrication is one of them. It is most essential. This is not only minimized tare and bare of the mechanical part of the antenna structure, motors, bearings, and gears but also bring down electrical consumption by motors. Lubrication makes moving part of the antenna working smoothly. To lubricate antenna following item required.

1. Cotton waste.
2. Recommended grease.
3. Grease Gun with rigid and flexible delivery tubes.
4. Kerosene oil
5. Spare nipples strait and bend both the types.
6. Watch maker's screwdriver set.
7. Open end wrench
8. Slide wrench

9. WD-40.
10. Isopropyl alcohol
11. Bristle soft brush nylon
12. Brstle hard brush nylon

Procedure:

1. Clean nipples with cotton waste. Scratch and remove any hard grease present on nipple. Inspect nipple for rust etc. Clean the nipples with alcohol soaked clothe and allow drying.
2. Touch the nipple's ball with Watchmaker screwdriver and see if spring action is there. If it is hard to push the ball, then better change the nipple with correct size and shaped nipple.
3. Nipples to be changed once noticed rust on it.
4. Fit the grease cartridge into the grease gun. Attach proper delivery tube to grease gun.
5. Connect delivery tube to nipple and tight it.
6. Pump the grease into nipple till fresh grease has come out from bearing and joint. Once fresh grease out stop feeding grease to the nipple.
7. Remove deliver tube from nipple. Wipe out excess grease from the joint and near the nipple.
8. Apply cap to the nipple to stop dust deposit and rusting.

Note: Locate all other nipples, refer antenna supplier's diagram for lubrication, and clean them as described above and grease them all. This completed antenna's greasing part of pressure greasing.

Replacement of nipple.

If nipple is not accepting grease you have to remove it and install fresh nipple with correct type. After cleaning the nipple area, as described above, spray WD-40 to the nipple and wait for few minutes. Recommended 5-10 minutes. Now use open-end wrench of correct size and fit it to the base of nipple and turn anti clockwise slowly. If you feel it is hard, apply WD-40 again and then turn to remove the nipple. Once nipple has been loose wiped-out area carefully before removing the nipple.

Remove the nipple and install new nipple by turning clockwise with hand first and then tighten it with wrench. Do not forget to get correct size and shape of nipple. Such as straight or bend type nipple to be replaced with similar types of nipple.

Greasing of Jackscrews:

Although jackscrew is greased by nipple provided for this purpose. The better way you should lubricate it and inspect screw condition also. This will help keeping jackscrew in right condition for future services. If you want to move your services to another satellite, then Jackscrew's position will change from present. To accommodate it you must keep the total jackscrew serviced, lubricated and free from rust.

Procedure:

1. Remove the rubber boot clamp near the motor- worm gear side.
2. Pull the booting way from joints at least take near another end (fixed end) of jackscrew and tie it properly.
3. Clean the jackscrew with cotton waste by removing hard grease on it.
4. Apply kerosene-soaked cloth and clean the jackscrew.
5. Make inspection for rust, tare, and bear of the jackscrew.
6. Check for smoothness of the groves if it is not smooth work on it with zero grade sandpaper.
7. Clean the jack screw completely
8. Allow for few minutes to dry before you apply grease over it.
9. Now apply thin coat of grease all over the jackscrew. Do not leave any portion without greased.
10. Apply second coat of the grease near worm gear end of jackscrew.
11. Carefully pull down the rubber boot so that grease should not wipe off.
12. Clamp and tighten the boot to the jackscrew housing.
13. Apply Denso tape to avoid clamp rusting.

This completed lubrication of Jackscrew.

Note:

Once you complete greasing of the Elevation jackscrew do the same for azimuth jackscrew too.

Warning: *While working on antenna, keep antenna-tracking unit on standby mode. Do not keep in Auto mode. Otherwise, motor will run, and you get hurt.*

Changing the Gear oil:

Motor gear assembly is filled with special oil. This is depending upon motor manufacturer and its type. Some motors use mineral gear oil. Adhere to manufacturer instruction how often we need to change the gear oil.

SEW motors comes with mineral gear oil type. The oil must be changed once in two years for mineral type and yearly for other type of gear oil.

Procedure:
1. Take an empty pot to collect old gear oil.
2. Open the bent hole plug with L and key.
3. Carefully remove the bent plug while keeping emptying pot below the hole.
4. Collect all the oil from gearbox.
5. Measure the old collected oil and record it (It would be Approx.250 ml)
6. Take similar amount of fresh gear oil, 250 ml or little more say for example 260 ml,
7. Close the bent hole with plug properly.
8. Open the leading oil hole plug with L and key.
9. Keep funnel on the hole and slowly pour the oil carefully in to funnel so that gear oil filled in.
10. Plug the hole properly and wipe out spilled oil from the gear assembly if any.

11. Record the date of filling gear oil in the maintenance dairy for future reference.
12. Keep checking gear oil by dipping screwdriver if level is going low top it little bit and check for leakage of oil.

Note:

Carry out above process for other motors gear assembly too. This completed gear assembly oiling.

Working on Rusted portion of the Antenna structure.

Antenna rusted portion must be noted while inspection of the antenna is made and rust to be removed and panting to be done. This is important to keep antenna structure strong enough and longer lasting. This is taken under annual antenna maintenance process.

Procedure:

1. Clean antenna structures with water and removes all dirt and salt deposits. You may need medium pressure water jet for this purpose. Otherwise, you can do it in normal way apply water and brush (iron brush) it.

2. Using Iron brush chirps all rust and remove it. Some time you may need special iron brush, which goes with drill machine. This will help you removing rust at odd places. They come in circular shape and different sizes to fit in our requirement.

3. Once all rust is removed clean the portion with water.

4. Apply rust removing chemical and allow 30 to 40 minutes. Wash the chemical with water clean the portion again with clean water and allow to dry.

5. Check again if all rust removed and antenna is rust free.

6. Apply antirust paint one coat and allow to dry it.

7. Finally apply cold Galvanise paint and allow to dry it.

Note:

Do similar work for all the rusted portion of the antenna and apply cold galvanize paint. Finally apply cold galvanize paint all over the antenna support.

Do not apply antirust chemicals to bearing, resolvers and other sensitive areas.

This completed antenna rust removing & painting process.

WARNING:

Take care of the power points, motor cubicles, and other delicate things if you are using water jet for cleaning. Protect them suitably by plastic cover.

Checking of the motors:

Annually we should check antenna motors cooling fan and brakes. To check it does followings:

1. Keep ACU in standby or manual mode.
2. Open the fan cover backside the motor.
3. Clean the fan by blower and then with cotton rags.

4. Check fan, tight and blades if lose ok.
5. Clean the brake assembly and see if it free from foreign material such as dirt, rust, cotton threads etc.
6. Check the brake by manually rotating the brake shoe by the rod provided for this purpose and see sufficient gap is there or not.
7. You can measure it by filler gauge and compare it by manufacturers data. If gap has increased, then brake may not be effectively working. In such case change the brake pad as described in the manual.
8. Carry out the above to other motor also. This completes the servicing.

Note: Changing of brake pad is beyond the scope of this book. Reason is every manufacturer has its own way of braking. So, it is advised to consult the manufacturer manual for this purpose. or consult expert in doing this.

Maintenance on Ground Communication Equipment

To day we have very sophisticated equipments in our earth station chain. VLSI has made it easy, slim and maintenance free equipments are available in the market.

Practically there is no need of maintenance on Modem, modulator, upconverters, down converters, LNA/LNB, Encoders and Multiplexers. To ensure proper operation of these equipment you should make some measurements monthly and calibrate the equipment yearly. Since most of the equipments are software based, to protect programs in the equipment a battery is installed inside the equipment. The life of battery, as per manufacturer, is 10 years. So once battery has gone low then you need to change the battery. You must check cooling fan and if not working properly then you must change that.

Maintenance on Modem:

1. Clean the modem with soft cloths.
2. Check cabling and connectors and tighten it if loose.
3. Measure DC voltages such as +/-12.0V, 5.0V, etc
4. Measure IF frequency and power in dB and record it.
5. Check cooling fan if giving abnormal sound get them changed. There is no lubricating facility on cooling recent types of fan.
6. Push the AC power card connected to unit properly.

Note: In case of problem with any cards inside modem it will generate an alarm and then you should change the faulty card. This is part of break down maintenance.

Maintenance on UP Converter and Down converter

Do following maintenance on maintenance free up converters.

1. Clean up converter chassis with soft cloth for dust and dirt.

2. Check cables and connecters and tight them properly.
3. Measure DC voltages as available at convenient points and see if they are within the limit otherwise adjust it.
4. Observe sound from cooling fan and in case of abnormal sound change the fan.
5. Measure and record followings:
    a. Reference oscillator frequency, level and its harmonics with level and signal to noise ratio.
    b. Phase lock loop (PLL) oscillator frequency, level and harmonics with level and S/N ratio
    c. RF frequency, its level and signal to noise ratio.

Note: You can plot with the help of plotter all above measurements for future reference. Or save in SA and make soft copy.

Maintenance on High Power Amplifier (HPA):

Here once again we have good HPA available which are maintenance free but still some maintenance they also need. You must remove unit from Rack to carry out maintenance.

1. Clean HPA with soft cotton cloth.
2. Open cover and check its cable and connectors for burn, loose connections, and bad connectors.
3. Check circuit modules, PCB for break, dirt deposits, itched out track, push buttons etc

4. Clean the modules with vacuum cleaner, blow air and brush. Do not use alcohol or any other solvent.
5. Clean housing inside with vacuum cleaner.
6. Check and clean air intake fan and exhaust fan.
7. Check and clean hot air blower
8. Check HPA connectors and push to tight it.
9. Measure all voltages and do necessary adjustment if required.
10. Close the HPA housing, installed on rack, and power it up.
11. Check its gain with min, middle and max power input. Record it. See this is giving good gain.
12. Make the RF measurements and signal to noise ratio. Plot the measurements by plotter.
13. Check for harmonics and its level too.

Note:

1. DO NOT WORK ON HPA IF YOU ARE ALONE.

*2. ADHER ALL WARNINGS BOFORE DOING ANY THING IN **HPA**.*

3. A LATHAL VILTAGE IS PRESENT INSIDE HPA ALL POSSIBLE PRECAUTIONS TO BE TAKEN WHILE WORKING INSIDE HPA AND WHILE MEASURING HIGH VOLTAGES.

SWITCHING ON & OFF HPA.

HPA to be switched on at 1st time or any time as follow:

1. Switch circuit breaker On in the HPA power supply unit. This will start prime power to HPA unit on. Self test initiated automatically. Power LED will glow.
2. If any problem persists Alarm or warning LED will come up.
3. After a delay of 5 minutes, HPA standby LED will come up.
4. Now you can press Beam On switch so that power can reach to amplifier unit.
5. Now you can see on RF unit of HPA input, reflect and out put window will display some reading.
6. Adjust input attenuator for correct power out from HPA and lock it.

Power Off:
 To switch off HPA do following simply do not turn CB "OFF".
1. Press beam Off to remove the power from RF unit
2. Left the unit at least 30 to 45 minutes so that TWT and another component in RF unit get cooled.
3. Note that blower/ centrifugal fan inside the RF unit still "ON" when beam Power is switched off.
4. After elapse of the time as in item (2) above, Switched off CB at HPA power unit.
5. Wait another 10 to 15 minutes if you want to work on HPA so that all capacitors get discharged in P.S.

6. Remove prime power cable from the HPA power unit to avoid accidental switch "ON" of the HPA unit.
7. Now it is safe to work on the HPA.

NOTE: It you switch off CB in HPA power unit accidentally then switch it ON again and do normal switch off procedure. This will allow TWT to cool inside and produce condensed water in the RF unit which when dripped to Power supply unit will cause major problem to power supply and Hence HPA.

Maintenance on LNA and Hub

You have not to do any maintenance on LNA/LNB. You must check: -

1. Coaxial and wave-guide switches are working properly or not.
2. LNA is getting proper power supply or not.
3. Feed system and LNA/ LNB assembly in the hub is ok or rust and dirt is deposited.
4. Clean any dirt and rust present inside hub.
5. Give touch up painting to the wave guide and mounting plate if rusted, paint came off etc.
6. Check for cable and connectors and tighten it.
7. Check heater for hub is working ok (this is provided if you earth station is in high humid area or rain remains round the year.)
8. Clean air intake filter in hub.

Chapter- 6

In this chapter, you will learn
1. Maintenance on Diesel generator.
2. UPS.

After completion of this you will be able to maintain power plant and diesel generator. Diesel generator is standby source of power for your earth station. You need un - interrupted power supply for your installation for continuous working in case of commercial power fails.

Maintenance of Diesel Generator

Diesel generator is a vital piece of our equipment, which takes active part once commercial power, fails. This allows our system to be **On- the- air** without break in service or interruptions. Once power failure is sensed by control circuitry it sends signal to Generator. Generator starts automatically. To start Generator Automatically we must ensure followings-

1. Diesel is in the tank.
2. Batteries are in good condition to supply the required current to start Generator.
3. Cooling system is working fine to keep Generator cool enough to work properly.

The above can be achieved by routine maintenance of the Generator.
Routine maintenance is divided into

A. Weekly,
B. Monthly
C. Annually and
D. Emergency maintenance any time.

Weekly maintenance.

Weekly maintenance on Diesel Generator is observations only. These are as follow: -

1. Check Battery water level, topped up if water level has gone down. In maintenance, free battery it is not require.

2. Check Battery charging system working properly and keeping battery charged or not. To do it just press boost charge button and in few seconds in should back to normal charge from boost. This indicates that battery is charged. Do not need any more charge.

3. Check radiator water level and add water if it is needed.

4. Check fan belt is ok. Press belt by finger you can feel tension. It should not be too tight or lose. If it is too tight is will create problem to generator. In this condition, Generator will start but take lots of current and starting will not be smooth. In case of marginal battery some time it may fail to start also. If it is too lose then cooling problem may occur. Engine will not get cooled because fan is not working properly even some time do not work if belt is loose.

4. Check Diesel level in the tank and keep its record.

Monthly Maintenance

Monthly maintenance is functional check. It should be carried out in two ways.

1. Run the Generator off load for 10-15 minutes and record the performance. Such as engine speed, voltage, frequency, normal operation, system alarm for under speed, over speed, engine hot, under voltage, over voltage etc.

2. Run the Generator onload by shout down the commercial power and record the performance of the generator as given in item –1 above.

The 1st method is used if GCE **is not** fully Auto redundant and cannot take care of un-interrupted transmission. The reason is once commercial power is cut or turned off load will go to battery of UPS and then generator will start, and load is transferred to the generator. During this changing process, there will be little jerks for a fraction of second. So, the GCE equipment might feel uncomfortable and try to switch to stand by one online. If standby equipment is not available there may be interruption in service/transmission.

The 2nd method is used if the GCE is in fully auto redundant mode of operation. This is preferred method, and you can get actual load current, voltage, and frequency of the generator power supply. Note these values in book for record.

Annual maintenance

The annual maintenance of the Generator is quite important. In this following maintenance is done.

1. Change the Engine oil
2. Change the Oil and water filter,
3. Drain the radiator water fill the fresh water with coolant in case if used.
4. Check the fan belt for tear and wear. Worn out belt must be replaced.
5. Diesel tank should be cleaned
6. Check electrical cables and other harness for lose and burn etc and correct it.
7. Cleaning, rust removing, and painting of the generator should be carried out.

Engine oil should be changed once yearly with right type of the oil as specified by the manufacturer.

Oil filter also to be changed annually otherwise the new engine oil gets contaminated once it came in circulation through old and dirty filter.

Radiator water must be change and fresh and clean water to be filled. Coolant should be used if specify by the manufacturer. Hose to be check for tight clamps and leakage etc.

Belt's tension to be checked, according to specification and in case worn out belt found then replacement should be made with correct types and size of the belt.

Depending upon the run of the engine diesel tank should be cleaned (If run is very often). In the case of rare run of the engine this step to be skipped.

Electrical cables, which are tightening by screw or nut – bolt, should be checked and if any lose connection found, tightened it. In case of burn cable ends which caused due to lose connections it should be repaired or replaced with new and correct size and rating of the cable piece.

Because of the environmental surroundings, cooling, and heating of the engine due to run of the engine, the body of the generator gets rusted. Rust should be removed /cleaned, and engine should be painted to avoid further damage of the engine.

Emergency Maintenance

This is most unfortunate and very rare type of the maintenance. No one expect this but some time it occurs even if weekly, monthly maintenance is done properly. Your engine is running properly and in between fan belt breaks and engine stops giving 'Engine Hot' alarm.

Some time when commercial power fails diesel generator could not back it because of the battery's sudden discharge and unable to start generator. In these cases if you have spare battery in the stock you can change. If you have spare fan belt you can change but the survival rate of the communication equipment is depending upon the battery backup duration of UPS and technical efficiency of the person attending to the fault.

This fault turns on alarm of technical person's and he works in tension, so an outage is expected. It can be saved only by two ways - First if you are lucky, commercial power returns before your backup battery exhausted.

And second your backup battery it self- if it is designed to take up load for longer time and Tech fix the problem within this time and able to start the generator to take load. Cross finger this type of the fault should not come.

Un- interrupted Power Supply (UPS)

Un-interrupted Power Supply system allows system run properly in case commercial power supply fails due to some reason or others. Normally power supply to the Telecommunication/ TV department is given on essential supply line. This is line where failure is rare. Generally, backup is given to this feed from commercial power Supply Company. In case of emergency, back up given from local generator. Once commercial supply fails total load transfers on the battery bank for few minutes (or few seconds as per adjustment for Diesel Generator to comes "ON"). Later diesel generator takes full load and battery goes on float charge. Generator back up power till the time commercial power remains unavailable to the installation.

Once commercial power resumes, after five-minute generator goes off. Thus, no interruption takes place in the communication system.
Once you need to attend ups due to some problem in it you can do manual bypass. In this case phase power, will be transferred directly to the load by passing UPS.
Now you can do servicing on the UPS either on converter or inverter section.

UPS should be checked every month by shut down commercial power and see battery backs up the load or not. Diesel Generator comes on after how many minutes of the commercial power gone off. Keeps its record for future use. You must clean the bent fan for dirt and dust with damped cloth.

Some UPS comes with more features in these you will get always AC load power from converter and in case any power fluctuation to the input load will go on bypass mode. These, UPS need proper alignment for phase and voltage out put. Very little variation is given for this purpose. Hence adjustment became very critical.

Annual Maintenance

You must perform following maintenance to UPS system.

1. Keep load on bypass mode.
2. Switch CB off
3. Take out Battery fuses so that inverter should not get any DC power.
4. Open the UPS system cover
5. Check for cable burn or connector burn.
6. Check PCB condition and clean it with blower.
7. Check other harnesses on the PCB and inside UPS.
8. Check bent fan and clean it.
9. Check Battery bank total DC supply and record it.
10. Check total current of the battery and record it.
11. Open battery cover and check individual battery conditions.
12. Replace bulged battery and faulty battery, which is not supplying correct current. (If some of the batteries are faulty or dead then reading in item 9 and 10 will not be correct as per specifications)
13. Clean battery bank and close the cover.
14. Put the battery fuse and switch on CB and measure AC out put voltage and phase. Do necessary adjustment to get correct recommended voltage and phase.
15. Switch off CB and normalise bypass switch and see how long battery supply the current to load.
16. A buzzer will sound when battery is going to nearly exhaust.

17. Run D.G. by switching off mains supply. Note down run-down time of the battery.
18. Check voltage and phase and frequency on the load side of UPS with Generator supply.
19. Switch on mains AC and switch off generator.
20. Measure AC voltage and phase on load side on UPS with commercial power supply and record it.

This completes UPS maintenance.

NOTE: You must keep recommended UPS spare such as Fuses, PCBs, and harness for connector.

Chapter-7

Brief information on satellite system

We are talking about satellite which is useful for communication, Radio and Television Broadcasting. This satellite is situated at 38000 kilo meter above the earth on equator. As we go away from equator its distance also increases from your location. It may be 40,000 kilo meters. Satellites are geostationary and their locations are fixed. The useful life of satellite is fifteen years. Proper station keeping is not possible beyond this period. Hence satellite goes in inclined orbit operation for few years before it is taken out of service. Satellites are monitor and controlled from ground Telemetry Tracking Control station periodically and controlled by issuing command if found drift from its correct location.

Command issued will direct satellite to move left or right, up or down to bring its correct location. Satellite always kept at centre of the box and should not cross the box limit. Fuel on satellites use to operator motors to move satellite in required direction. When this fuel is nearing to finish in approx. fifteen years, satellite station keeping become critical and satellite declared to be inclined orbit satellite. It is positioned weekly to its location. This satellite moves in letter "8" path in one direction. So weekly commanded to move in other direction. Satellite operator issue Ephemeris data for its inclined orbit satellite to all the users so that they can feed in their antenna controller to track the satellite effectively.

Beacon signals are available from all satellite and operators are using this through their beacon receiver and controlling their antenna so that all the time it looks the satellite and establish good communication or TV broadcasting.

Beacon frequency can be had from satellite operator. They may have two sets of beacons one in Horizontal and other in Vertical polarization. Operators uses generally this beacon which is in their service pol. Some satellite has four beacon signals two in each pol. Operator may use as they feel comfortable. Some beacon receivers have two in put and can be selected as per requirement. Hence two beacons can be accommodated in it.

We know now we have satellites in C- Band, Ku-band, DBS- band and Ka- band. These are serving for voice, Data and TV services. Services on other than C-band other bands are get affected during rain at site. Now satellites are coming with more transponders. Transponders are 36 MHz wide. Some satellite has transponders 72 MHz wide. This can accommodate more services on a transponder. When SCPC services are there on transponder vigil to keep on it so that no suppression occurs in any service due to power up by adjacent service. MCPC occupy full transponder of satellite and mux more services on it.

Telecom generally uses 2 to 4 MHz for one service. If they have more carriers then multiple of 2 or 4 MHz is used.

Chapter- 8

Miscellaneous items and tools:

You need proper tools to do proper work. The installation company supplies tool an earth station needs. In case it is not awarded in contract then you must have your own set. I am listing here necessary tools and measuring test equipments you need more frequently.

Test Equipment:
1. Spectrum Analyser
2. Power meter with power head sensor
3. Frequency counter.
4. Power head
5. BER test set.
6. Multimeter –Digital Meter (Fluke)
7. Noise test set
8. Test translator.
9. Plotter
10. Network analyser
11. RF signal Generator

Note: installer supplies most of the above test equipments. It is lifetime equipment for earth station.

Tools:
1. Screwdriver set.
2. Box spanner set
3. Open end wrench set
4. Hammer

5.   Files different size & types.
6.   Hand drill electrical low power.
7.   Vacuum cleaner with blower facility.
8.   Grease gun with fixed and flexible delivery tube.

Note: All items listed above required to buy <u>once only</u> for the lifetime of the station.

Consumable spares:

This is spare which you consume while doing maintenance. Once you will carry out a specific maintenance then you will find our exact amount of the things you need for whole the year to carry out maintenance.

1. Grease cartridge for greasing antenna.
2. Anti seize grease
3. Gear oil
4. Galvanize paint
5. Solvent for above
6. Black paint spray type 1 can
7. Blue paint spray type 1 can
8. Distilled water
9. Oil filter
10. Water oil separator filter
11. Fan belt appropriate type
12. Engine Oil
13. Coolant for radiator
14. Silicon sealant.
15. WD-40
16. Anti rust spray

17. Anti rust paint.
18. Grease nipples straight and 45 deg bent.
19. Iron brush straight and circular different size
20. Hairbrush 1", 2" size
21. Nylon brush 1"
22. Cotton rags
23. Paper towel
24. Soft cotton towels

Note: All above items are consumable and for **ONCE** only use. You have order for a year. I guess you need 4x above items for whole the year maintenance use. This is recurring expenditure. Items should be as per manufacturers recommendations.

APPENDIX -A

Schedule for earth station installation.

        This is an example only. Based on the fact you may make your own schedule. It is in fact made to suit customer's requirements. Customers want his earth station run as quick as possible. So, the installation company must make the proposals and the schedule to suit requirement of the customer. How ever some margin for contingency must be provisioned.

Work \Month	Jan	Feb	Mar	Apr
Initial Meeting	1w			
Servey team visit	3 w			
Pemilinary proposal		1w		
Final proposal		3 w		
Contract Signing			2 w	

B. Shipment of Materials

        Once contract is awarded to company and signed by both the parties (customer and company) company will go for shipment of the material to the site.

Materials \ Month	May	Jun	Jul	Aug
Foundation materials & documents	2w			
Structure & PD antenna		1w		
GCE & Test equipments		3w		

C. Work completion schedule

Once materials arrived at site the work should start as per contract. Some work to be done by customer and then further company will carry out till completion, testing and commissioning.

Works \ Months	Jul	Aug	Sept	Oct
Antenna foundation & Equipment room	1w			
PAD Antenna assembly	3-4 w			
GCE rack with equipment		1~3w		
In station testing / staff training		1~3w		
Testing and commissioning			1~2w	
Handover to customer			3w	
Spares and documents handover				1w

Note: 1. w: stands for week and preceding figure are that week of the month.

Appendix- B

In station test and final test and commissioning

In station test must be carried out as per Satellite System Operation Guide issued by satellite operators/ Organizations in modules. For example, SSOG- 210, issued by INTELSAT explains this test and schedules

In station test must be completed well before final test and commission is implemented. Sometimes equipment behaves very funny such as level variation of HPA or upconverter, frequency changes, ACU problem. So, in station test make sure that this funny behaviour is taken care of.

Request is sent from customer for permission from satellite operator to allot transponder, frequency, band width and test schedules for up link their carrier.

Satellite operator then issue:

1. Date of start and end of the test.
2. Time of the start of the test.
3. Name and contact number for Communication system monitor (it is an earth station having facilities to down link any satellite in the zone and make measurement for frequency and EIRP etc)

4. Test results to be submitted to satellite operator after completion of the test. Test result should be forwarded to your correspondent too.
5. Correspondent earth station will also forward his observations or test results to satellite operator copy to you.
6. CSM will forward his results to satellite operator only.
7. All results to be submitted on prescribed form only.

Certification:

After getting test results from you, and **CSM**, satellite operator makes study of the result and will insure that accepting your services from the earth station antenna under test will cause no harm to any other existing or future services through his satellite. Once study finished the satellite operator issue your earth station certification. Now you are certified to operator the services from your earth station.

Service Operations:

Once you received certificate from satellite operator, you and your correspondent earth station should issue service request to the satellite operator this includes.
1. Date of service request.
2. Start date of the service.
3. End date of the service.
4. Type of the service- voice/data/video
5. Number of the service.

6.   Bandwidth utilized by this service- 2, 2*X, MHz [where X =2,3,4 etc] or 36 MHz for video services.
7.   Customer's signature.

Commencement of the service:

Practically you can start the service once test is over and satellite operator say go head on phone. Officially satellite operator replies your request and in that they will send their agreement and date from which you will be charge for the services with standard published rate of satellite service provider. You have to consider this date as operational date. Satellite operator gives few days of grace period. Intelsat generally gives 15 days grace period to their customer

Appendix-C

Earth Station Information

The blow data is just as example need to be completed by you from your actual station data.

Transmit

Parameter	Serve –1	Service-2	Service-3
Transponder No.	37/ GA	37/ GA	11/ WH
Frequency (MHz)	6374.9400	6369.4575	5997.1075
Up Cont. C.F.	6063.00	6063.00	5965.00
Modem IF (MHz)	154.9400	146.4575	172.1075

Receive

Parameter	Serve –1	Service-2	Service-3
Transponder No.	15 /WH	15/ WH	37/GA
Freq. In MHz	4058.7125	4051.1088	4150.7275
Dn Conv C.F.	4055.00	4055.00	4135.00
Modem IF MHz	143.7125	136.1088	155.7275

Earth station Capabilities:

Operating:
Transmit polarisation- LHCP/RHCP or   A/B   Both

Receive polarisations- RHCP/LHCP or   B/A   Both

ACU Tracking capability -Can track in
1.Auto mode
2.Program track mode, and
3.Manual mode.

These are the things you must have readily available for your Earth Station. This will help you in understanding transmission and reception plans. It will also help you when you are complaining any problem to satellite operation centre. They need this information from you to make correct observations.

ACU DATA:

This is vital data for your information. Please keep record of all the applicable parameter for your earth station ACU so that in case of emergency you can track satellite manually. Please make a note for elevation, azimuth, and beacon level for each of the satellite you are transmitting signal.

Azimuth: 174.5
Elevation: 34.5
Polarization: 75
Beacon level: 0.90 dB. As an example. (you must note down for your earth station operating on each satellite)

This will help you if your ACU failed to track in auto mode and you lost the satellite. So, manually you can bring elevation and Azimuth angle display same as recorded above to get antenna back in action. This is also helpful to earth station operator when he changes a new ACU for his earth station once the working one goes faulty or its battery exhausted and replacement has been made.

If you are operating in stationary orbit satellite, then Depending upon ACU model and make different parameter needed for ACU is explained in the book.

Giving one manufacturer example would be injustice. I am leaving this to station operators to fetch this information from ACU and keep a record of DATA in the Earth station.

Beacon frequencies:

Satellite operators are transmitting number of unmodulated carriers, called beacon, of small band width which is received by receiving earth station for tracking of satellite. This frequency is in band. Generally given in the guard band of the transponder. Satellite operators **must** notify all their customers about their beacon frequency and its periodicity i.e. which beacon frequency is available and when. Generally, they transmit at least two beacon frequencies in a band. Here is list of INTELSAT beacons for C-band and Ku band.

Beacon No.	C-Band	Status	Ku-Band	Status
1.	3947.5 MHz	ON	11.198 GHz	ON
2.	3952.5 MHz	ON	11.452 GHz	OFF
3.	3950.0 MHz	ON (Vertical/ Linear)		

Thus, an earth station technician can select any one beacon for tracking of the satellite. Based on your beacon receiver capability and linearity you must select beacon. If you display these beacons on spectrum analyser you will find its level and band width. Bandwidth is same for all beacons. Some beacons have little high level than other. It may be 0.5 to 1 dB high.

Appendix-D

<u>Earth Station GCE Rack Lay out</u>.

It is also advisable to keep earth station equipment rack layout ready in the earth station. Though all the equipment is levelled but it will give you details which equipment used for what carrier. Hence in case of problem any one reach at correct piece of equipment. It is seen in a panic tech generally disturbed working equipment considering this equipment is part of faulty chain. Thus, he shut down another link too. Here I am giving details of the rack layout as an example.

Earth Station Equipment room or shelter which ever you are using must be fitted with cooling system such as window A/c or Central A/c depend upon your requirement. If you have more TV Transmission uplink system from earth station and rack mount HPAs, central a/c with high power needed. It should be with backup facility.

Please see below sheet for details of Rack lay out.

GCE Equipment in Rack for [A] country Telecom link

PATCH PANNEL	
Up con-1	FOR [A] COUNTRY CXR
Switch	
Up con-2	BU
Down con-1	FOR [A] COUNTRY CXR
Switch	
Down con-2	BU
IF mon RF	FOR [A] COUNTRY CXR
BU Modem	BU FOR BOTH CXR'S
Modem switch	
Modem-1	FOR [A] COUNTRY
Modem-2	FOR [B] COUNTRY
Blank	
Blank	
Fan	FOR COOLING

NOTE :

Right side box, against each equipment, shows for which carrier it is used.

GCE Equipment in Rack for [B] country Telecom link

PATCH PANNEL	
Up con-1	FOR [B] COUNTRY CXR
Switch	
Up con-2	BU
Down con-1	FOR [B] COUNTRY CXR
Switch	
Down con-2	BU
IF mon RF	FOR [B] COUNTRY CXR
Blank	
Blank	
Fan	FOR COOLING

NOTE :

Right side box, against each equipment, shows for which carrier it is used.

GCE Equipment in Rack-TV Transmission DTH link

Rack-1

LNA TEST PANEL
L-BAND Mon
EIRP MON
Blank
TLT H-Pol   TLT V-pol
TLT Patch panel
Blank
Up Converter BU
Up converter 1
Up converter 2
Blank
DBS Patch Panel
Blank
Blank
Up converter-BU
Up converter 1
Up converter 2
Blank
Modulator-1
Modulator Switch
Modulator-2
Blank
Modulator-1
Modulator switch
Modulator-2
Serial Hub
Moxa 5016

GCE Equipment in Rack-TV Transmission DTH link

Rack-2

RF Patch Panel
Blank
HPA-1
Blank
HPA-2
Blank
HPA-BU
HPA Controller
Blank
H- band 1:2   BDC Chassis
Blank
Low Band 1:2   BDC chassis
BDC controller
Blank
L- Band patch panel
Blank
Blank

Antenna installation process:

Digging for foundation of antenna PAD

Pouring concreate for foundation

71

Assembling reflector of antenna

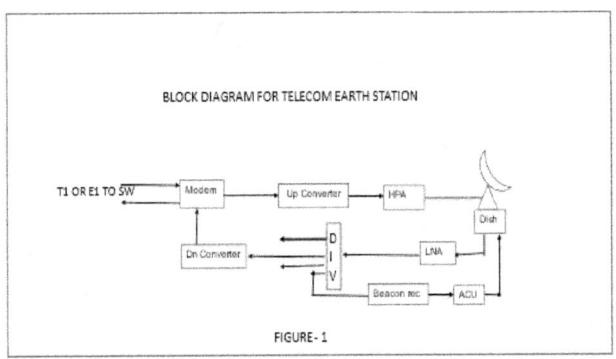

FIGURE- 1

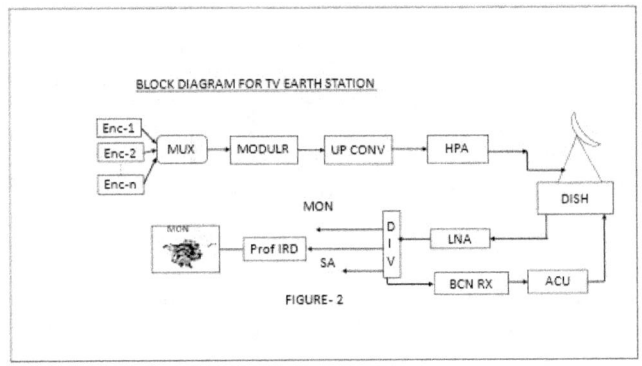

FIGURE- 2

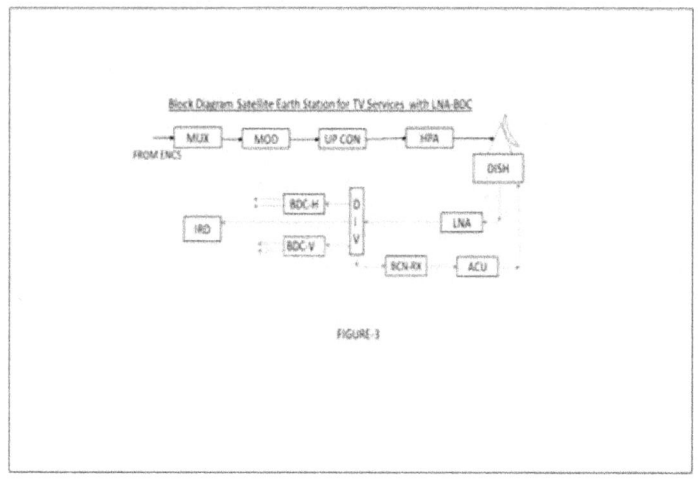

Block Diagram Satellite Earth Station for TV Services with LNA-BDC

FIGURE-3

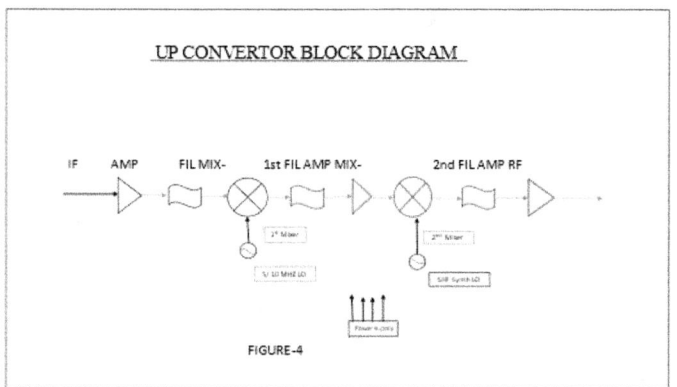

UP CONVERTOR BLOCK DIAGRAM

FIGURE-4

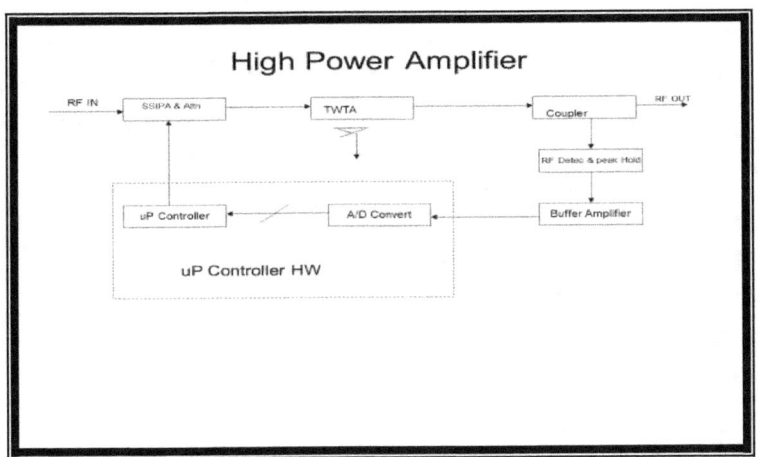

Figure-5

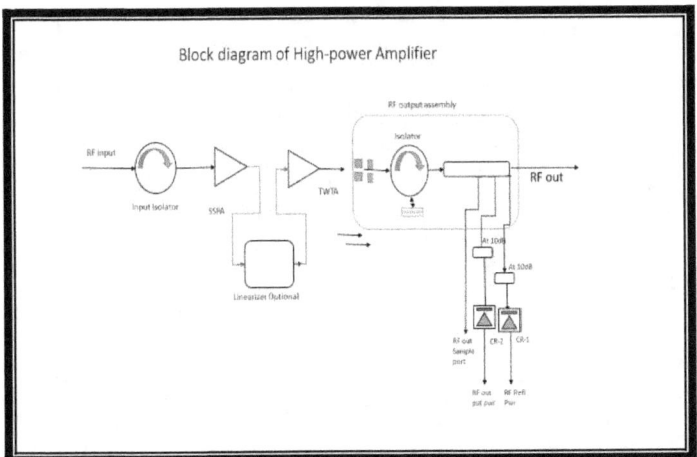

Figure- 6

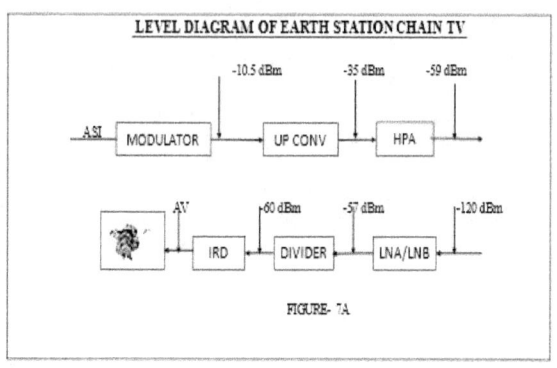

LEVEL DIAGRAM OF EARTH STATION CHAIN TV

FIGURE- 7A

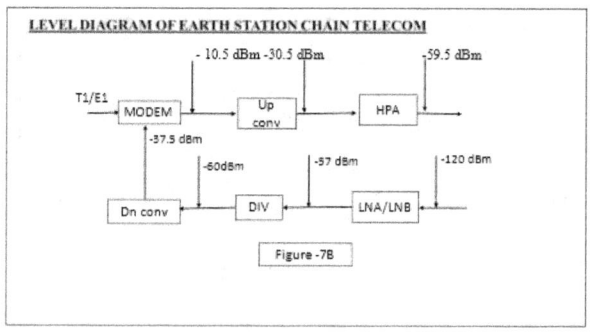

LEVEL DIAGRAM OF EARTH STATION CHAIN TELECOM

Figure -7B

76

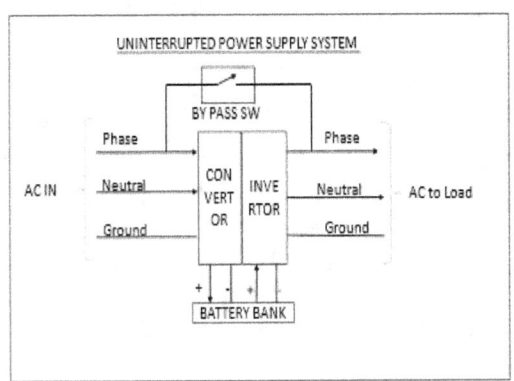

HIGH POWER AMPLIFIER performance										
Month			Fortnight				HPA			
DATE	TIME	O/P PWR	REF Pw	ATT set	HLX_C	HLX_V	Htr hr	BM_hr	Temp	RM T	REMS

Table-1

HIGH POWER AMPLIFIER performance											

Month			Fortnight			HPA					

DATE	TIME	O/P PWR	REF Pw	ATT set	HLX_C	HLX_V	Htr hr	BM_hr	Temp	RM T	REMS

Table-2

www.ingramcontent.com/pod-product-compliance
Lightning Source LLC
Chambersburg PA
CBHW071031220526
45467CB00004B/1612